Field Notebook

Observe, Record, and Reflect on the Natural World

Imbo Lohman

Idle Winter Press
Portland, Oregon

Idle Winter Press
Portland, Oregon
http://IdleWinter.com

First published 2022
Printed in the United States of America

The text of this book is in Baskerville

ISBN-13: 978-0615891538 (Idle Winter Press)
ISBN-10: 0615891535

Table of Contents

Table of Contents 3
4
5
6
7

How to Use This Book 8
Field Guide 9
10
11
12
13
14
15
16
17
18
19
20
21
22
23
24
25
26
27
28
29

30
31
32
33
34
35
36
37
38
39
40
41
42
43
44
45
46
47
48
49
50
51
52
53
54
55
56
57
58
59
60

Data Collection 61
62
63
64
65
66
67
68
69
70
71
72
73
74
75
76
77
78
79
80
81
82
83
84
85
86
87
88
89
90
91

92
93
94
95
96
97
98
99
100
101
102
103
104
105
106
107
108
109
110
111
112
113
114
115
116
117
118
119
120
121
122

123
124
125
126
127
128
129
130
131
132
133
134
Completed Investigations 135
136
137
138
139
140
141
142
143
144
145
146
147
148
149
150
151
152
153

How to Use This Book

The ***Table of Contents*** will help you keep track of what you have put into this book. Each time you use a page, be sure to add the title of your page to the corresponding page number in the table of contents, so you can find it again later.

In the ***Field Guide*** section, devote each page to a different organism. Draw or attach a photograph of it in the large square, and then fill out the rest of the page as best you can. Feel free to use books and other resources to add useful information. Make sure your entry is complete enough to help you identify the organism in the future.

The ***Data Collection*** pages are designed to be versatile. Choose blank pages, lined pages, graph paper, or the best combination for whatever you need to record. Don't forget to write the date on your work!

Finally, the ***Completed Investigations*** section is a way to keep track of your accomplishments. Each time you complete an investigation, place a sticker, drawing, or title in one of the squares, and write the date below the square.

The 8-inch ruler on the ***back cover*** is useful for measuring specimens.

Field Guide

draw or attach specimen

Common Name:

Scientific Name:

Date Found:

Location:

Size:

Observations:

draw or attach specimen

Common Name:

Scientific Name:

Date Found:

Location:

Size:

Observations:

draw or attach specimen

Common Name:

Scientific Name:

Date Found:

Location:

Size:

Observations:

draw or attach specimen

Common Name:

Scientific Name:

Date Found:

Location:

Size:

Observations:

draw or attach specimen

Common Name:

Scientific Name:

Date Found:

Location:

Size:

Observations:

draw or attach specimen

Common Name:

Scientific Name:

Date Found:

Location:

Size:

Observations:

draw or attach specimen

Common Name:

Scientific Name:

Date Found:

Location:

Size:

Observations:

draw or attach specimen

Common Name:

Scientific Name:

Date Found:

Location:

Size:

Observations:

draw or attach specimen

Common Name:

Scientific Name:

Date Found:

Location:

Size:

Observations:

draw or attach specimen

Common Name:

Scientific Name:

Date Found:

Location:

Size:

Observations:

draw or attach specimen

Common Name:

Scientific Name:

Date Found:

Location:

Size:

Observations:

draw or attach specimen

Common Name:

Scientific Name:

Date Found:

Location:

Size:

Observations:

draw or attach specimen

Common Name:

Scientific Name:

Date Found:

Location:

Size:

Observations:

draw or attach specimen

Common Name:

Scientific Name:

Date Found:

Location:

Size:

Observations:

draw or attach specimen

Common Name:

Scientific Name:

Date Found:

Location:

Size:

Observations:

draw or attach specimen

Common Name:

Scientific Name:

Date Found:

Location:

Size:

Observations:

draw or attach specimen

Common Name:

Scientific Name:

Date Found:

Location:

Size:

Observations:

draw or attach specimen

Common Name:

Scientific Name:

Date Found:

Location:

Size:

Observations:

draw or attach specimen

Common Name:

Scientific Name:

Date Found:

Location:

Size:

Observations:

draw or attach specimen

Common Name:

Scientific Name:

Date Found:

Location:

Size:

Observations:

draw or attach specimen

Common Name:

Scientific Name:

Date Found:

Location:

Size:

Observations:

draw or attach specimen

Common Name:

Scientific Name:

Date Found:

Location:

Size:

Observations:

draw or attach specimen

Common Name:

Scientific Name:

Date Found:

Location:

Size:

Observations:

draw or attach specimen

Common Name:

Scientific Name:

Date Found:

Location:

Size:

Observations:

draw or attach specimen

Common Name:

Scientific Name:

Date Found:

Location:

Size:

Observations:

draw or attach specimen

Common Name:

Scientific Name:

Date Found:

Location:

Size:

Observations:

draw or attach specimen

Common Name:

Scientific Name:

Date Found:

Location:

Size:

Observations:

draw or attach specimen

Common Name:

Scientific Name:

Date Found:

Location:

Size:

Observations:

draw or attach specimen

Common Name:

Scientific Name:

Date Found:

Location:

Size:

Observations:

draw or attach specimen

Common Name:

Scientific Name:

Date Found:

Location:

Size:

Observations:

draw or attach specimen

Common Name:

Scientific Name:

Date Found:

Location:

Size:

Observations:

draw or attach specimen

Common Name:

Scientific Name:

Date Found:

Location:

Size:

Observations:

draw or attach specimen

Common Name:

Scientific Name:

Date Found:

Location:

Size:

Observations:

draw or attach specimen

Common Name:

Scientific Name:

Date Found:

Location:

Size:

Observations:

draw or attach specimen

Common Name:

Scientific Name:

Date Found:

Location:

Size:

Observations:

draw or attach specimen

Common Name:

Scientific Name:

Date Found:

Location:

Size:

Observations:

draw or attach specimen

Common Name:

Scientific Name:

Date Found:

Location:

Size:

Observations:

draw or attach specimen

Common Name:

Scientific Name:

Date Found:

Location:

Size:

Observations:

draw or attach specimen

Common Name:

Scientific Name:

Date Found:

Location:

Size:

Observations:

draw or attach specimen

Common Name:

Scientific Name:

Date Found:

Location:

Size:

Observations:

draw or attach specimen

Common Name:

Scientific Name:

Date Found:

Location:

Size:

Observations:

draw or attach specimen

Common Name:

Scientific Name:

Date Found:

Location:

Size:

Observations:

draw or attach specimen

Common Name:

Scientific Name:

Date Found:

Location:

Size:

Observations:

draw or attach specimen

Common Name:

Scientific Name:

Date Found:

Location:

Size:

Observations:

draw or attach specimen

Common Name:

Scientific Name:

Date Found:

Location:

Size:

Observations:

draw or attach specimen

Common Name:

Scientific Name:

Date Found:

Location:

Size:

Observations:

draw or attach specimen

Common Name:

Scientific Name:

Date Found:

Location:

Size:

Observations:

draw or attach specimen

Common Name:

Scientific Name:

Date Found:

Location:

Size:

Observations:

draw or attach specimen

Common Name:

Scientific Name:

Date Found:

Location:

Size:

Observations:

draw or attach specimen

Common Name:

Scientific Name:

Date Found:

Location:

Size:

Observations:

Data Collection

Completed Investigations

Idle
Winter
Press

Made in the USA
Monee, IL
04 November 2026

67382356R00085